MORLEY LIBRARY

3 0112 1003 3714 3

W9-AIA-158

14.50

MORLEY LIBRARY
184 PHELPS STREET
PAINESVILLE, OHIO 44077
(440)352-3383

Everyday Science Experiments in the Gym

John Daniel Hartzog

The Rosen Publishing Group's
PowerKids Press™
New York

Some of the experiments in this book were designed for a child to do together with an adult.

Published in 2000 by The Rosen Publishing Group, Inc.
29 East 21st Street, New York, NY 10010

Copyright © 2000 by The Rosen Publishing Group, Inc.

All rights reserved. No part of this book may be reproduced in any form without permission in writing from the publisher, except by a reviewer.

Photo Illustrations by Shalhevet Moshe

First Edition

Book Design: Michael de Guzman

Hartzog, John Daniel.
     Everyday science experiments in the gym / by John Daniel Hartzog.
          p.   cm. — (Science surprises)
     Includes index.
     Summary: Suggests activities and experiments that demonstrate the principles of science at work in a school gym, exploring such topics as gravity, friction, and oxygen.
     ISBN 0-8239-5458-7 (lib. bdg.)
     1. Science—Experiments Juvenile literature. [1. Science—Experiments. 2. Experiments.] I. Title. II. Series: Hartzog, John Daniel. Science surprises.
     Q164.H2744   1999
     507.8—dc21   99-18287
                                                                                              CIP

Manufactured in the United States of America

Contents

14.50

Secrets of the Gym

We all know that the gym is a place to discover what your body can do. What you might not know, though, is that by making your body try new and different things, you are also learning about science. Everything you do in the gym depends on science, whether it's throwing a ball to a friend, learning to move in a new way, or even stopping to catch your breath. Science is all around you, and it's even inside you. These **experiments** will show you how science works in the gym.

◀ *The games you play in the gym can help you learn about how your body works.*

Playing Catch

When you throw a ball to a friend, do you ever aim straight at them? Probably not, because if you do, **gravity**, the **force** that pulls all things toward the center of the earth, will pull the ball down to the ground long before it ever reaches your friend. Instead, most people throw the ball up high, so it will come back down as it reaches the other person. That's because a ball moves in a curve. It goes up when you throw it, and then gravity pulls it back down again. Try this experiment.

Materials Needed:
- Any ball

When you throw a ball, it always travels in a curve.

Watch two friends throw a ball back and forth. Have them try to throw the ball all different ways, including, high, low, straight, or curved. You will see that no matter how they throw it, the ball always travels in a curve. The science of how balls travel can help us in the gym.

Materials Needed:
- 1 pair of sneakers
- 1 pair of socks
- 1 pair of dress shoes

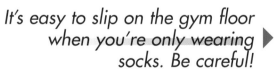

It's easy to slip on the gym floor when you're only wearing socks. Be careful! ▶

Friction

If you ever played a game of basketball in your socks, you probably slid right across the floor. You need sneakers to keep you from falling. The soles of your sneakers are usually made of **rubber**. The rubber creates **friction** between your feet and the floor. Friction is the force that resists movement between two surfaces that touch each other. Take off your shoes and put on a fresh pair of socks. Try to run a little on the gym floor and then stop. Chances are you'll keep sliding. Your socks don't create friction on the smooth gym floor. Now try it with a pair of shoes that are not sneakers. Dress shoes have smoother soles than sneakers, so they create less friction. Imagine what would happen without friction in the gym!

Listen to Your Heart

Sometimes you play so hard in the gym that you have to sit down and catch your breath. That's because of some of the science that's happening

Materials Needed:
- 1 cardboard toilet paper tube
- A friend

inside your body. Your body is working so hard at running and playing that it needs extra **oxygen** and energy. The heart is an **organ** that helps to deliver this energy to other parts of your body. It works like a pump, pushing blood with plenty of oxygen in it out to the rest of your body. You can hear your heart working with this experiment.

Put the toilet paper tube up to your friend's heart and listen. Can you hear his heartbeat? ▶

Before you start doing anything in the gym, put one end of the toilet paper tube to your ear and the other end against a friend's chest. You can hear the beating of his heart. It should make a pounding sound, like *badoom-badoom*. Have your friend run around for a little while, and then listen again. Is his heart making louder noises? Are the noises happening faster? As his body works harder, so does his heart.

Muscle Memory

You've probably noticed that the things you do in the gym, like obstacle courses or even pull-ups, get easier the more you practice. That's because each time you do an exercise, your muscles remember a little more about how it's done. To see muscle memory in action, try this experiment.

Lean one side of your body against a strong wall. Using all your strength, try to lift your arm straight out into the wall. Count to 25 slowly while you keep trying to lift your arm with one long push. Step away from the wall and gently try to place your arm straight down at your side. Your arm should

Your arm will keep trying to lift itself, even after you stop pushing! ▶

keep floating up again. It is still trying to lift straight out, even though your brain tells it to stay still. The muscles in your arm remember the earlier message so well that they are still trying!

The Bounce in the Ball

Materials Needed:
- 1 basketball
- 1 tennis ball

Basketballs and tennis balls both bounce. Can you guess which bounce higher? You can find out the answer and the reason why through this science experiment.

Hold the tennis ball in one hand and the basketball in the other hand at equal heights from the ground. The basketball will bounce higher because it is heavier. Its weight gives it more energy to bounce. Now let's try something a little different. Hold the tennis ball just above the basketball and drop them together. The basketball will hit the floor and the

◀ *First drop the two balls next to each other. Which bounces higher?*

tennis ball will also bounce off the top of the basketball. The tennis ball bounced a lot higher than before because it gained energy from the basketball's bounce. Next, try dropping the basketball on the tennis ball. The basketball does not bounce very high because the tennis ball doesn't have much bounce to give it.

The tennis ball goes a lot higher when it bounces off the basketball. ▶

Materials Needed:
- Tape
- Some string
- A tennis ball

Throwing a Curve Ball

People have always wondered how pitchers throw curve balls. Pitchers do this by spinning the ball with their hand as they throw it. The spin makes the ball change direction. This experiment will show you how spinning affects the curve a ball makes.

Tape a piece of string onto a tennis ball. Hold the string in one hand and let the ball hang right in front of you. With your other hand, twist the tennis ball about 50 times. Let the ball go and watch it spin. The spin of the ball begins to push the ball through the air. When a pitcher throws a curve ball, the ball spins and changes its direction. It still goes toward the batter but it also curves.

◀ *Watching a ball spin can help you understand how throwing a curve ball works.*

Shooting Hoops

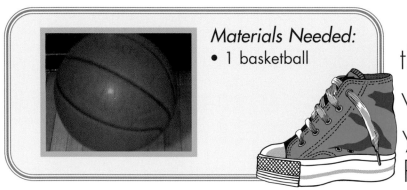

Materials Needed:
- 1 basketball

Even though you might think that your arms do all the work when you shoot a basketball, your legs are working, too. Pushing with your legs helps push the ball closer to the hoop. If you are not sure how important your legs are in shooting a basketball, find a spot where the shots you make usually hit the rim. Now sit down right where you are and try the same shot. Were you able to hit the rim that time? Without your legs, you probably couldn't. Your whole body works in most sports. Most times your body's parts work together without you telling them to.

Your legs help you throw the ball into the basket.

Can you make the same shot sitting down?

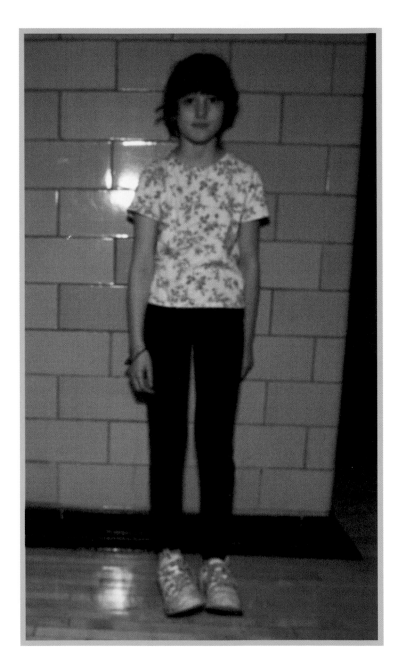

Can You Balance Without Your Toes?

When you are standing, running, or playing sports, you use every part of your body to be **balanced**. To be balanced is to be in a steady position. Try this experiment to see how important both feet and all your toes are to keeping your balance.

Stand with your side pressed up against a wall. Get as close to the wall as you can while your feet stay flat on the ground. Once you're in this position, try lifting your outside foot without moving away from the wall. Can you keep standing without this foot? Now try standing as straight as you can with your back against the wall. Try to lift your toes off the floor. Did you know how important your toes are to keep you standing up straight?

◀ *Get as close as you can to the wall, and see if you can move your feet without losing your balance.*

The Science of the Gym

Who knew you could learn science while playing in the gym? Athletes can play better if they learn more about their sports, and the scientific laws that control their bodies, their equipment, and their games. Many of these laws are described by a type of science called **physics**. Physics helps us understand movement, forces, matter, and energy. It helps answer questions like, "Why do sneakers help us play basketball?" and "What makes balls bounce?" Science is everywhere, even in the gym!

Glossary

balanced (BAL-uhnst) In a steady condition or position.

experiments (ehks-PER-ih-ments) Tests that are used to find out something or the effect of something.

force (FORS) Something that moves or pushes on something. Force can change the direction or speed of a moving object.

friction (FRIK-shun) The rubbing of one thing against another. The force that resists movement between two surfaces that touch each other.

gravity (GRA-vih-tee) The force that pulls things downward towards the earth.

organ (OR-guhn) A part of an animal or plant that is made of special tissues and that does a particular job.

oxygen (OKS-ih-jen) An invisible, odorless gas that living things need to stay alive.

physics (FIH-ziks) The scientific study of matter and energy, and the laws that govern them. Physics is the study of motion, force, light, heat, sound, and electricity.

rubber (RUH-ber) A strong elastic, waterproof material that comes from the sap of a certain tropical tree.

Index

A
athletes, 22

B
balance, 21

E
energy, 10, 14, 15
exercise, 12

F
force, 6

friction, 9

G
gravity, 6

H
heart, 10, 11

M
muscles, 12, 13

O
oxygen, 10

P
physics, 22

S
sneakers, 9, 22
spinning, 17

Web Sites:

You can learn more about science in the gym on this Web site:
http://www.kidnaround.com/science.html